Examining Energy

Examining

Solar Energy

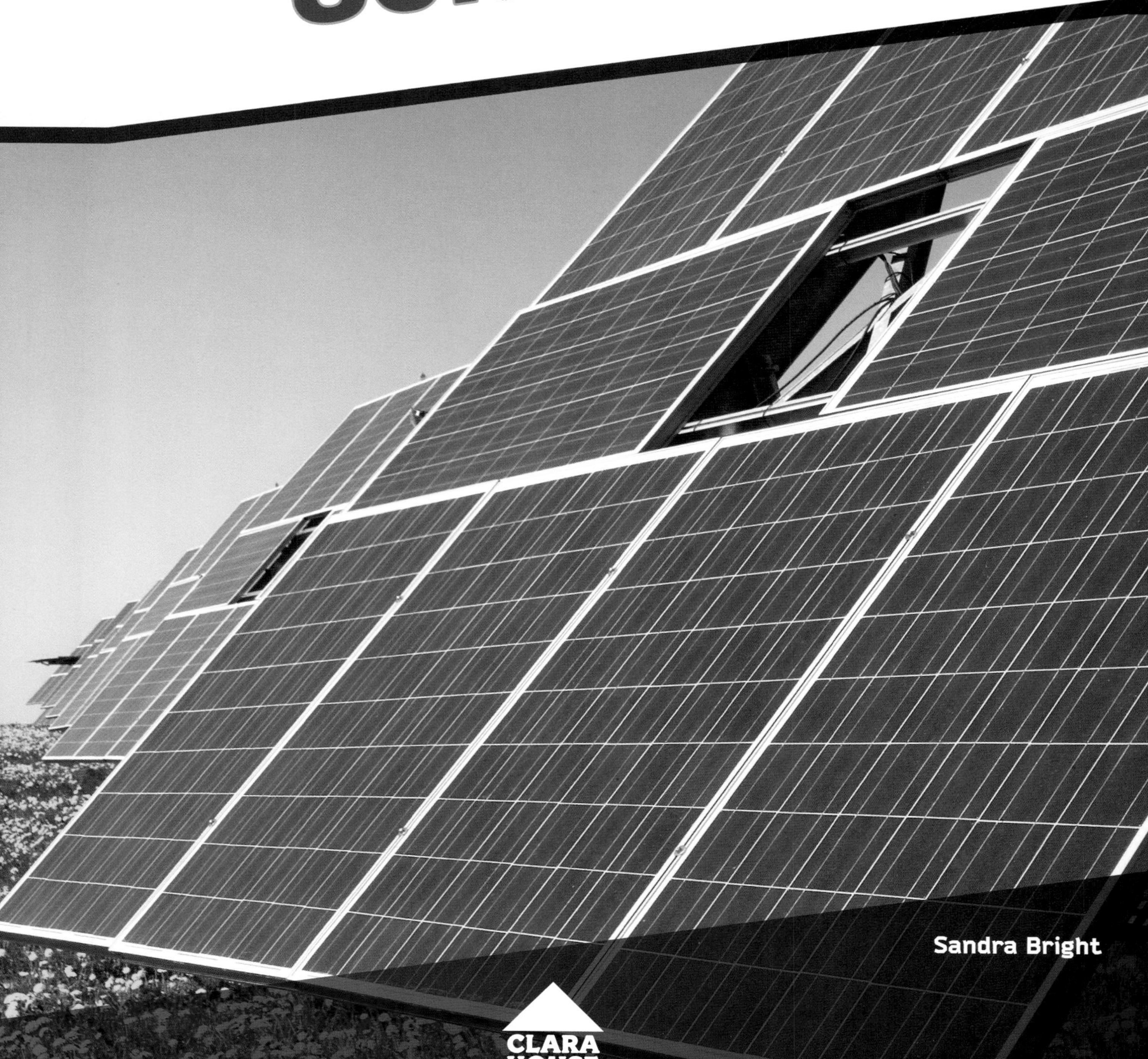

Sandra Bright

CLARA HOUSE BOOKS

First published in 2013 by Clara House Books, an imprint of The Oliver Press, Inc.

Clara House Books
5707 West 36th Street
Minneapolis, MN 55416
USA

Produced by Red Line Editorial

The publisher would like to thank Dr. Barry D. Bruce, Professor of Biochemistry, Cellular & Molecular Biology & the Microbiology Departments for the University of Tennessee at Knoxville, for serving as a content consultant for this book.

Picture Credits
Fotolia, cover, 1, 19, 30, 41; Shutterstock Images, 5, 14, 20; Lazar Mahai-Bogdan/Shutterstock Images, 8; ARENA Creative/Shutterstock Images, 9; Vitoriano Jr./Shutterstock Images, 11; Uryadnikov Sergey/Shutterstock Images, 13; NASA, 23; Andy Z./Shutterstock Images, 24-25; Tom Grundy/Shutterstock Images, 27; Red Line Editorial, 29; OConnell/Shutterstock Images, 33; Glenn Campbell/AP Images, 34; Kurhan/Shutterstock Images, 37; Andrey Bayda/Shutterstock Images, 39; Blend Images/Shutterstock Images, 45

Library of Congress Cataloging-in-Publication Data
Bright, Sandra.
Examining solar energy / Sandra Bright.
pages cm. -- (Examining energy)
Audience: Grades 7 to 8.
Includes bibliographical references and index.
ISBN 978-1-934545-45-4 (alk. paper)
1. Solar energy--Juvenile literature. I. Title.
TJ810.3.B75 2013
621.47--dc23
2012035316

Printed in the United States of America
CGI012013

www.oliverpress.com

Contents

CHAPTER 1

Energy for the Future

Has anyone in your family ever complained about the high cost of electricity? Have you heard people talk about the cost of heating a home during the winter? Right now, a lot of the world's energy comes from non-renewable sources. These non-renewable sources, such as oil, are often bad for the environment, and they will eventually run out. Because of growing worldwide demand, energy keeps getting more expensive.

Right now, more than 80 percent of the energy that powers our modern way of life comes from oil, coal, and natural gas, which are all fossil fuels. These energy sources are made from organic materials that have been buried underground for millions of years. Because we are using them much faster than they can be created, these sources will eventually run out. Fossil fuels also give off a lot of carbon dioxide as they are used,

Solar energy is an important resource that is already helping to meet our energy needs.

which can lead to pollution. Scientists are constantly looking for ways to improve our energy sources. They want to find ways to produce energy that are more efficient, less expensive, and better for the environment than our current energy sources. Alternative energy research focuses on balancing our energy consumption with the needs of our environment.

One possible alternative source of energy might be the sun. Sunshine, or solar energy, is currently used to light homes, heat water, and produce electricity. Finding a way to expand our use of the sun's energy effectively could lower overall energy costs. Additionally, solar energy comes from a renewable source—the sun's energy will never run out.

Energy from the sun comes in the form of solar radiation and travels in electromagnetic waves. It takes about eight minutes and 20 seconds for solar radiation to travel from the sun's surface to the earth. Some of these electromagnetic waves arrive as light waves that humans can see. These light waves come in all the colors of the rainbow. But some electromagnetic radiation is invisible to us. People have been using the light of the sun for thousands of years. Expanding our ability to harness solar power may play a key role in the future of energy.

EXPLORING SOLAR ENERGY

In this book, your job is to learn about solar energy and its place in our future. How does sunshine light homes, heat water, and produce electricity? What technologies can help us use more solar power? Can sunshine replace other energy sources that are polluting the planet? How will solar power help shape our energy future?

Sage Cooper is researching solar power to help her school decide whether to add solar panels to its roof. She is meeting with scientists and other solar energy experts who will help her learn more about solar energy and the work it can do. Reading Sage's journal can help you with your own research.

CHAPTER 2

Solar Source

I know the sun is an energy powerhouse. Solar energy helps plants grow and powers the wind and the rain. Without the sun's energy, our planet would be just another lifeless rock in space. I'm visiting the Pennsylvania Science Center. Dr. Ben Rodriguez, a solar astronomer, has agreed to show me the sun through the center's special telescope. You should never look at the sun on your own—doing so can cause serious damage to your eyes.

Through the telescope's lenses, the sun looks just like the huge ball of hot gas that it is. Dr. Rodriguez explains that the sun is so big that more than 1 million Earths could fit inside of it. Its energy comes from nuclear explosions deep inside it. "The sun's core is about 27 million degrees Fahrenheit, or 15 million degrees Celsius," says Dr. Rodriguez. All that energy eventually heads out into space in electromagnetic waves. Most of the solar radiation reaching the earth is in the form of visible light—in all the colors of the rainbow.

Dr. Rodriguez takes me outside to the center's sunny rooftop patio, where he shows me another demonstration. Tall green grass is growing next to a black box.

Under the box are a few short, sickly, pale grass sprouts. "The grass under the box is the same kind of grass growing next to the box," Dr. Rodriguez says. "But without energy from sunlight, the grass in the box cannot make the food it needs to grow."

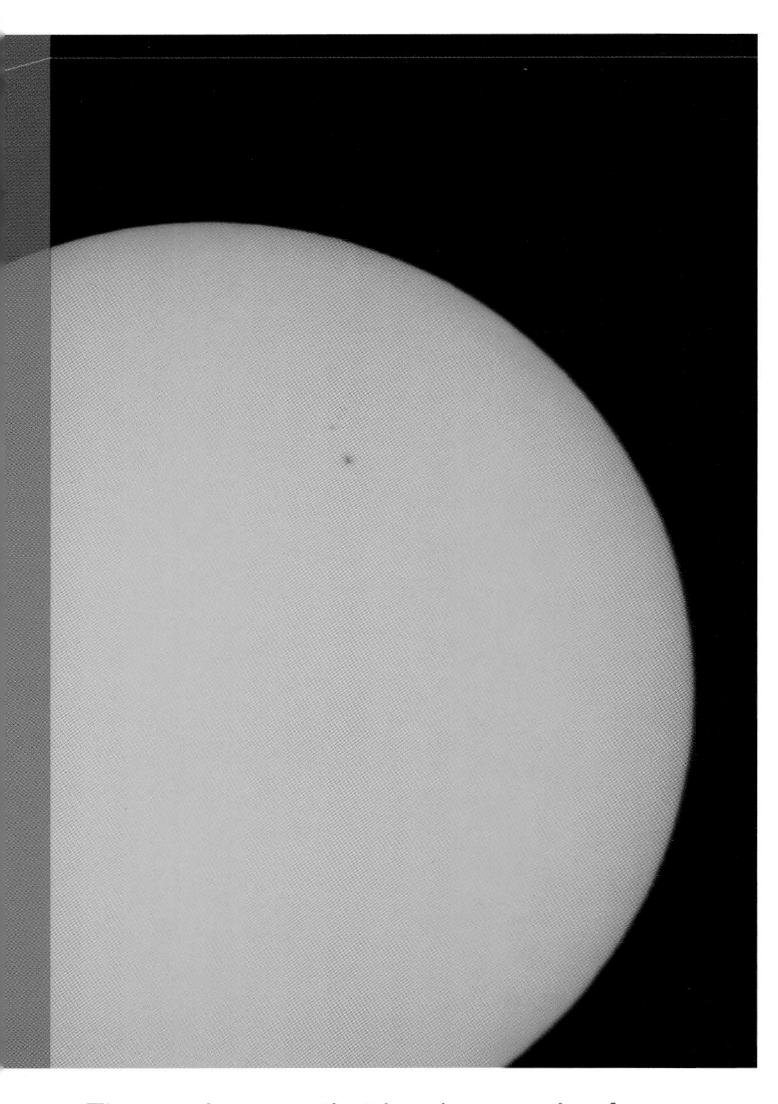

The sun is a star that has been active for more than 4.5 billion years.

He tells me that green plants have a light-absorbing pigment called chlorophyll. The energy that chlorophyll absorbs powers a chemical reaction between water and carbon dioxide. That reaction is called photosynthesis, and it's how green plants make energy. Photosynthesis makes carbon-based sugars that plants—and the animals that eat them—depend on to live.

Plants rely on energy from the sun to survive.

"Fossil fuels also really come from the sun," Dr. Rodriguez tells me. "They come from animals and plants that died long ago and, over a very long time, became buried deep underground. Heat and pressure from the overlying rocks turned them into carbon-rich energy sources we can burn to run factories,

ALTERNATIVE ENERGY TODAY

We don't have to wait to use some kinds of alternative energy. In fact, you may already be using them in your daily life. Check your calculator, watch, and outside lights. Do they have solar cells? Next time you're with a parent or other adult who is getting gas, check the pump. Does the label say "10% ethanol"? That would mean one tenth of your car's fuel comes from corn or other plants. Or the pump might read "20% biodiesel." Then your car may burn soybean or other plant oil!

heat homes, and power computers."

Dr. Rodriguez tells me that demand for fossil fuel is growing as the earth's human population grows. More people use more energy. Also, poor countries are industrializing—which means building factories and roads so their citizens can live like people in wealthier nations.

"Fossil fuels aren't like sunshine, either," Dr. Rodriguez says. "They are non-renewable resources. That means we have only so much oil, coal, and natural gas on Earth. Once they're used up, they're gone."

That's why it's important to find a way to use alternative energy sources on a large scale, Dr. Rodriguez tells me. Solar power, wind, flowing water, and biofuels are major sources of alternative renewable energy.

As I leave the science center, I think more about what Dr. Rodriguez told me. If fossil fuels are really running out, my school will need to find another way to produce its energy.

I understand that plants and animals need the sun to survive, but I'm still not sure how the sun's energy could be used to light a dark room.

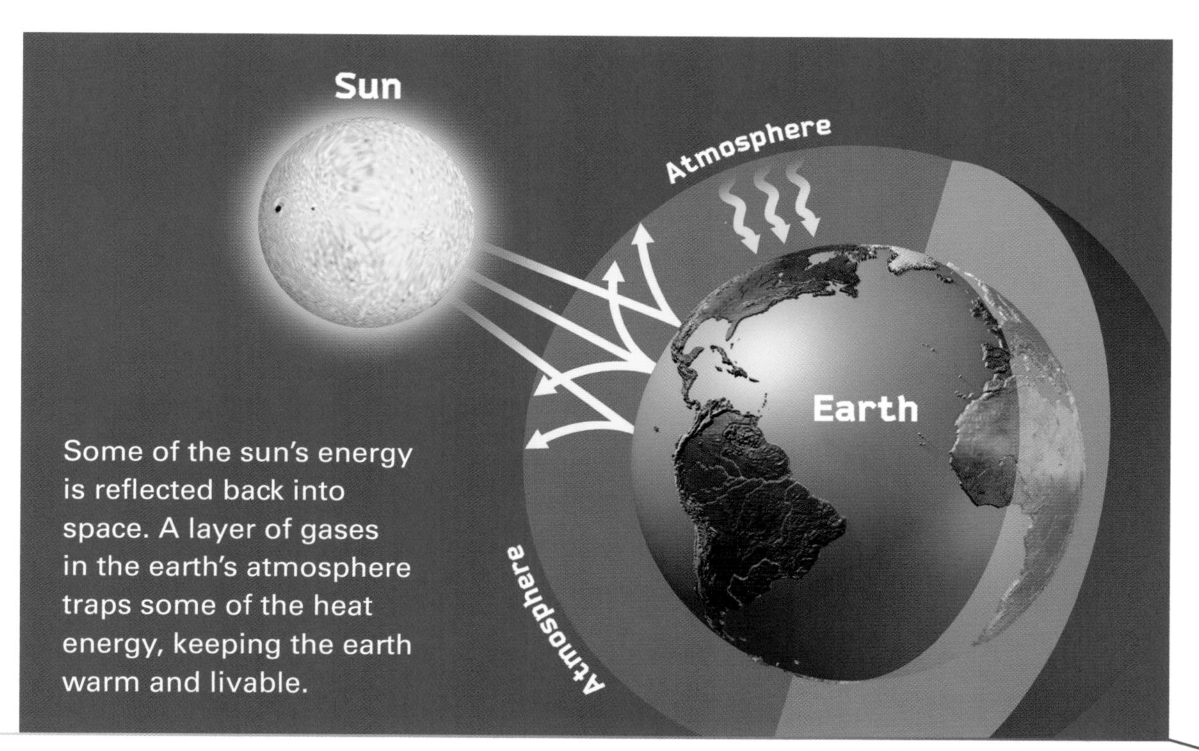

Some of the sun's energy is reflected back into space. A layer of gases in the earth's atmosphere traps some of the heat energy, keeping the earth warm and livable.

A CHANGING CLIMATE

Climate change is nothing new. Since the planet formed long ago, natural cycles and events such as volcanic eruptions have warmed and cooled the earth. In 1896, Swedish physicist Svante Arrhenius warned about something called the greenhouse effect. Arrhenius's research showed that too much carbon dioxide in the atmosphere could trap solar energy, heating the planet the same way glass traps heat inside a greenhouse. Some scientists believe the gases given off by burning fossil fuels are contributing to this greenhouse effect, causing more heat to be trapped inside the atmosphere. They think this extra heat is causing a gradual warming of the earth's average temperature.

CHAPTER 3

Heat and Light

"Ouch!" My bare hand touched a rock baking in the Arizona sun. "Feels like a frying pan!" I am visiting Beth Ortillo, an educator at High Creek Nature Center in Arizona. I want to learn more about how people use sunlight for energy.

"Ancient American Indians used hot stones to heat water," Beth tells me. "Stones have thermal mass, which means they can absorb heat and hold it. Today, we use thermal mass at the nature center to stay warm all winter."

Beth shows me around the nature center's patio. "See that lizard? He's absorbing sunlight, too. He needs the sun's rays to keep his body warm." She tells me that some of the sun's radiation is reflected into our eyes in the form of light, allowing us to see objects and colors. However, much of the sun's energy

Cold-blooded animals, such as lizards, need the sun to heat their bodies.

Adobe houses are a great absorber of the sun's heat.

is absorbed as heat. People can use the sun's heat to warm water or heat their homes—or a nature center.

Ancient Pueblo American Indians built their houses using a mixture of stone or mud bricks called adobe. Adobe acts like the rock I touched. It bakes in the sun all day, holding heat. When the air cools, the adobe radiates heat energy into the air, warming the building. The Pueblos knew the sun's path changes through the seasons. Because the sun is low in the sky during winter, they built their houses facing south in order to catch the most possible sunlight.

"We still use this style of heating today," Beth tells me. "The nature center's walls are made of adobe."

Beth invites me inside the nature center. A wall of south-facing windows and skylights brighten the center. "Clear glass lets the sun's radiation into the building," she explains. "Instead of paying for electric lamps, we have natural sunlight. This is called passive solar energy."

She tells me that solar buildings in colder places use sunlight to heat tile floors or columns of water that store extra heat. They may also have a greenhouse, where glass traps so much solar heat that tropical plants can grow—even during the winter. "Greenhouses have been around a long time, too," she adds. "President George Washington grew lemons in his greenhouse!"

"But if solar energy has been used for such a long time, why are we still so dependent on fossil fuels?" I ask.

"Because not every building is designed like the nature center," Beth explains. "Does your house have skylights and walls with thermal mass? It's difficult and expensive to add these things to existing buildings. Also, it won't work well in many places, such as a house in the northern woods."

Beth also reminds me that passive solar energy can't power a lot of our electric devices, such as lights, computers, and refrigerators. "The sun radiates more energy to Earth in an hour than humans use in a year," Beth says. "But to turn this energy into something we can use, we have to transform it into electricity."

CAN YOU TEST THERMAL MASS?

What's a good building material choice for passive solar buildings? You can test different materials in your oven to find out. Collect five cans and fill each one halfway with one of five different materials: water, sand, small stones, salt, and soil. With an adult's help, put the cans in an oven heated to 150 degrees Fahrenheit (65°C). After one hour, carefully remove the cans from the oven using protective oven mitts. Take the temperature of each can every five minutes and record your findings. Which material cools the fastest? Which cools the slowest? Which material would be the best choice as a building material for a passive solar house?

CHAPTER 4

Putting Sunshine to Work

Beth takes me outside to meet with High Creek Nature Center's naturalist, Rashi Thaker. Rashi is a desert biologist. She studies the adaptations that help plants survive in Arizona's heat and drought.

Rashi leads me along a path following a line of soaring metal towers. She tells me the lines are high-power transmission lines. They carry high-voltage electricity from a coal-burning power plant into town. Most of the town is connected to these wires. "The huge system of electrical lines is called the grid," says Rashi, "and everyone who uses it is 'on the grid.'"

"Are you on the grid?" I ask.

"Yes and no," says Rashi. She points to a small house near the path with solar panels on the roof. "This is my house. On sunny days, the panels make electricity that replaces some of the power I would get from the grid. I also dry my laundry in the sun and conserve energy whenever I can. Those are some of my energy adaptations," she says.

But with a computer, lights, air conditioning, and other appliances, she says the panels can't do the whole job. She saves money by reducing the amount of electricity she buys, but she is still on the grid.

A few more blocks of walking and we come upon a small grocery store. "This is my parents' store," says Rashi. "They're adapting to energy limits, too." The whole flat roof is covered with solar panels. "Take a look at this electric meter."

I notice a digital display marked kWh. From science class, I remember kWh stands for kilowatt-hour, a unit measuring how many kilowatts of power are used in an hour. For example, my laptop computer takes about .02 kWh to run for an hour. However, on this meter, the numbers are going backward.

DANGER LINES

Are there high-power transmission lines near you? More and more of these lines now snake through woods and grasslands as world electricity use grows and grows. They often have a big impact on the land. To put up the lines, utility companies cut down wildlife habitat. And to maintain easy access to the towers, companies may mow and use chemicals to kill off the plants that try to grow back. To wild animals, the land under transmission lines can be as hard to cross as a four-lane highway. Without plant cover, animals may be unable to find food or, worse, predators may catch too many of them.

Installing solar panels on your roof can save energy, but solar panels can be expensive.

Rashi says, “On most days, my parents’ solar panels make more energy than they need. My parents sell the extra power back to the utility company.”

Rashi tells me some people don’t even have the option of connecting to an electrical grid. A weather station on a mountain or an island lighthouse might not be reachable by electrical lines, for example. And poor countries often can’t afford to build transmission lines to reach remote villages.

About 1.4 billion people in the world live without electricity, she tells me. But that’s where solar panels can help.

Solar panels in remote places can help provide energy to people who may not be able to connect to the electrical grid.

With one small solar panel, a family can power a few light bulbs and a phone charger. They don't need to wait for a grid of transmission lines. They can improve their lives by adapting to solar energy now. There are problems with this, though. For one, solar panels are expensive. It can take a long time for a solar panel to save enough money to be worth the investment. Another problem is that the sun is not always shining. People want to use lights most often at night. They need to capture solar energy and store it to use when the sun is not shining. Batteries can do this, but the batteries are also expensive.

CHAPTER 5

Solar Electricity

I'm on the roof of the science building at a university in New Mexico meeting with Dr. Brenda Richards. The roof is covered with solar panels. I have seen solar panels from the ground before, but never this close up.

Dr. Richards points out that each panel is made up of dozens of smaller squares, or solar cells. That's where sunlight is transformed into electricity, she says. The cells are called PVs, or photovoltaics. "*Photo* means light," she tells me, "and *voltaic* means electricity."

The first PVs, built during the 1970s, were made of the element selenium. "Those early cells were less than 1 percent efficient," says Dr. Richards. "Imagine your homework was a set of math problems. If you were 100 percent efficient, you could finish in one hour. How long would it take if you were 1 percent efficient?"

Yikes! 100 hours! No wonder scientists kept working to try to create better PVs. During the 1950s, scientists found that they could use silicon, an element found in sand, to make a 4 percent efficient solar cell. By the 1960s, PVs connected into solar panels were efficient enough to power a research satellite orbiting the earth.

SOLAR POWER IN SPACE

Solar cells work even better in space than they do on Earth. In space, there are no clouds to block the sunshine. Scientists first used solar cells in space in 1958 on the satellite *Vanguard I*. This satellite, which was powered by six solar cells, collected and sent data back to Earth for six years. Today, many kinds of satellites orbit the earth powered by solar cells. Even though there are no clouds in space, some of the solar energy these satellites collect is still stored in batteries. On the *International Space Station*, winglike solar panels power operations, experiments, and life-support systems for a crew of six.

Dr. Richards explains that when the sunlight hits the silicon panels, bits of energy from the sun called photons cause the electrons in the silicon to move around. These electrons flow through wires in the panel, creating an electric current. Wires coming from the panels use this electric current to power lights and other building needs. The wires can also feed energy into a battery. This battery transforms electrical energy from the solar cells into chemical energy. Inside the battery, chemical energy can be stored to use later, when the sky is dark. “But batteries are expensive. And they can only store so much energy. This means that most people using solar panels will still have to connect to the electrical grid.

Energy from the sun powers the International Space Station as it orbits the earth.

"The panels work well for our science building, where we want to show people that clean, renewable energy is a possibility. But the panels are still less than 20 percent efficient, and they cost a lot to install. Coal, which is the fuel used most often to generate electricity, is more efficient."

"Doesn't the energy made by the solar panels help to decrease the money you spend on electricity?" I ask.

"Yes, but PVs are so expensive," says Dr. Richards. "It takes at least 10 years of energy savings to pay back their cost. It also takes a lot of energy and resources to create a solar panel. A solar panel needs to run for several years before it produces enough clean energy to offset the cost of the energy used to build it."

We use a lot of energy at night when the sun is not shining.

Dr. Richards has given me a lot to think about. Solar energy can work to power a single building. But it's not very efficient, and it can be very expensive. I wonder if there is a way for these solar panels to power more than just one building at a time.

CHAPTER 6

Powering Up

I'm at a farm in the California desert. The sign at the gate says "Golden Solar Farm," and from a distance the crop looks like acres and acres of mirrors.

Dave Gregg is the farm's chief engineer. "I just visited a new PV solar plant in Israel," Dave tells me. "It has 18,500 PV panels and produces 9 million kWh of electricity per year."

Dave tells me that his system is different. Instead of using photons, the Golden Solar Farm uses solar heat to create electricity. This is called concentrated solar power (CSP). I remember the big windows at the nature center. The windows captured the sun's heat to warm the building. This solar farm traps the same heat but uses it to create electricity.

We're close to the mirrors now, and I can see row after row. Each one is very long and shaped like an animal feed trough.

Solar farms are one way to produce solar energy on a large scale.

EARLY CSP

Huge solar farms using CSP may be a new concept, but CSP has actually been around for a long time. During the 1830s, a British astronomer named John Herschel made a solar oven using an insulated box. Reflectors inside the box concentrated the sun's heat in order to cook food. Ever since, inventors have been experimenting with ways to harness the sun's heat.

Dave points to the closest one. "See that tube down the middle?" he asks. "The tube is filled with oil. The curved mirrors catch the sunlight and focus it on the tubes. Sensors help the mirrors track the sun across the sky all day, so they heat the oil up. After the mirrors heat up the oil, the oil heats water in the tubes to a boil. The water becomes steam, which turns a turbine, and the turbine powers an electricity generator."

There are other kinds of solar plants, he adds. In some plants, thousands of flat mirrors surround solar power towers. Tracking the sun, the mirrors focus light on the tower, where the concentrated beams heat melted salt to over 1,000 degrees Fahrenheit (538°C). Melted salt works better than oil because it can get much hotter and hold heat longer, allowing the plant to operate even after the sun has set. "The rest of the process works like our system," explains Dave. "Water heated to steam turns a turbine that runs an electric generator."

We drive past more and more rows of collecting mirrors. "The CSP system works well, but like all energy options, it is imperfect," Dave says. "One problem is how much land solar farms require. As you can see, the mirrors take up a lot of space that used to be animal habitat."

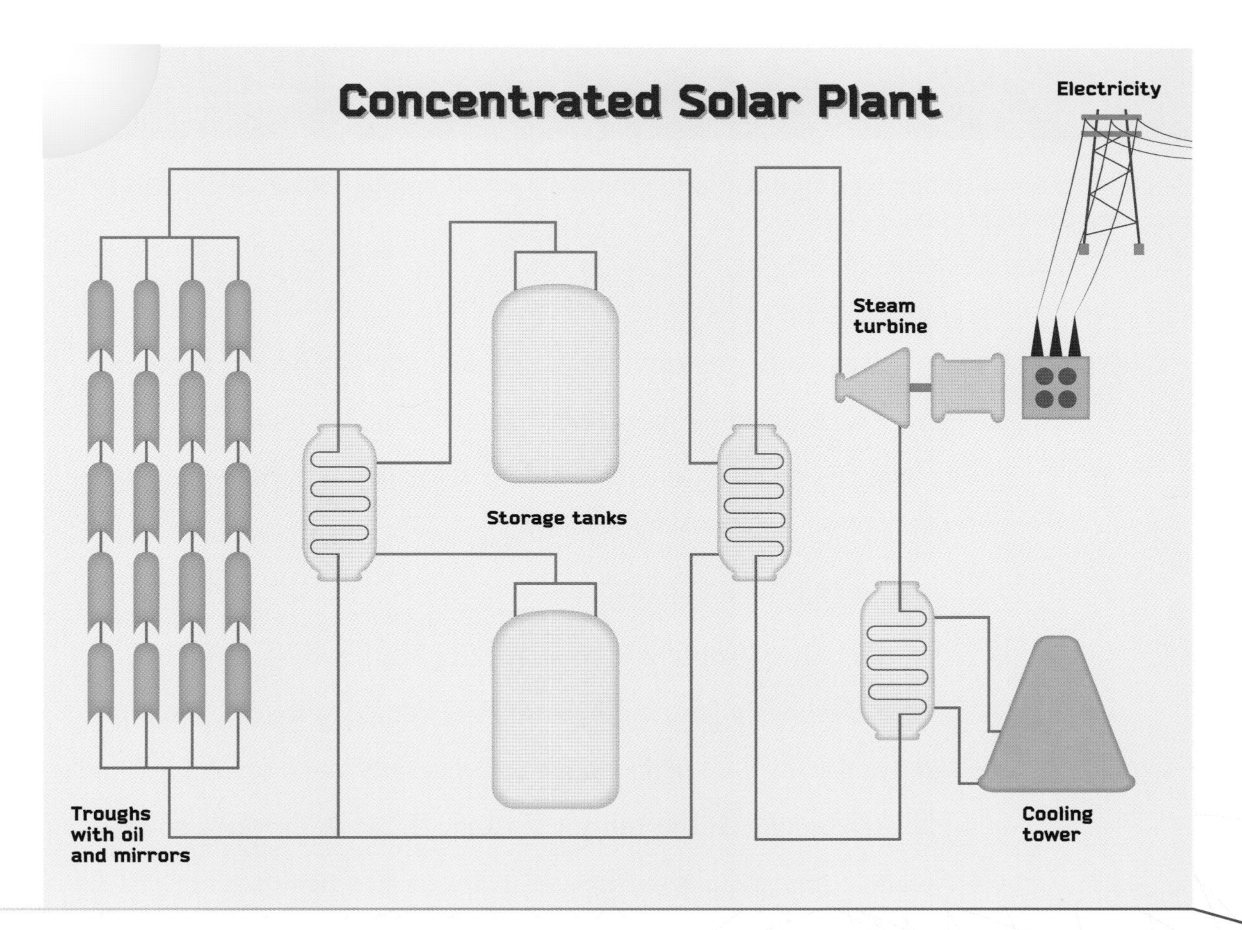

CONCENTRATED SOLAR PLANT

A lot of sunshine reaches the earth every day—but the sun's heat needs to be concentrated before it can be harnessed to power industrial work. The CSP method works by using the sun's heat to power a generator. The process takes heat energy from the sun, changes it into movement energy in the form of steam, and then changes that movement energy into mechanical energy in the form of a spinning turbine. The motion of that turbine is finally used to power a generator, transforming mechanical energy into electrical energy.

Solar farms are expensive to build and require a lot of space, which takes away from natural habitats.

Dave tells me territorial animals such as desert tortoises have the biggest problem with solar farms because they can't just move on to a new place when solar mirrors take over their habitat. Also, eagles, hawks, and other birds often die when they crash into long solar transmission lines.

Another problem is cost. All the land and equipment to make electricity from solar energy is very expensive. Dave tells me that fossil fuel prices have to be very high in order for a solar plant to be worth the money it takes to mass-produce solar power. Right now, solar power is about five times as expensive as electricity created by power plants that use fossil fuels.

"But remember," says Dave, "engineers are improving the technology every year. As it gets more efficient, plants will need less land and solar power will cost less."

Sunshine may be free, I think to myself, but solar energy isn't cheap. I wonder how solar power works in places that aren't as sunny as a California desert.

CHAPTER 7

Light Moves

I still have some questions about solar energy. I can see how it would work in a place famous for its sunny skies, like California, but I wonder how it could work in a rainy place like Washington. I also wonder if solar power could ever replace gasoline as a fuel for cars.

I'm meeting Tess Rush, the mayor of Solis, Washington, at her town's fair. Mayor Rush tells me that cloudy skies don't necessarily mean a place can't use solar power. Germany is a leader in solar power, and it gets even less sunshine than Solis.

"But our lack of sunshine does present challenges," the mayor says as she guides me to a fair exhibit featuring a demonstration of a PV panel. "The panel is still working," she tells me, "just not as efficiently as it does on a clear day."

Wires from the solar panel lead to a large battery. I remember from the university that the electrical current from the solar panel charges the chemical solution inside the battery, which can be used to make electricity. This battery is huge.

"There's a real drawback to solar power around here," Mayor Rush says. "We need to be able to store a lot of energy for the really cloudy days, and these big batteries take up tons of space. They also cost a lot of money."

Next the mayor leads me to a small racecourse. Eight solar-powered toy cars rush down the track, cheered on by the students who built them. The car that crosses the finish line first looks more like a wing on wheels than my family's car.

"Even with our solar power, the town still uses as much oil as other towns," she says. "We haven't found a practical way to convert solar power into the kind of liquid fuel that powers today's cars." The mayor tells me about a solar car race in Australia. People drive cars nearly 1,900 miles (3,000 km) across the Australian outback to prove that solar energy can

NEVER-ENDING SUNSHINE

One reason people like solar power is that the sun is reliable. The amount of energy reaching our atmosphere from the sun is almost constant: 1.361 kilowatts per square meter, or 1.361 kW/m2. However, the amount of sunshine available for work is not constant. The amount of energy reaching the earth can vary due to the weather, the time of day, the time of year, and even an area's geographic location. Because of the shape of the earth, the equator gets the most direct sunlight. The farther north or south a place is, the less energy it receives. Places that are very far north or south, such as Alaska or Antarctica, may get very little sunlight during some months and have very long days during other months.

Solar panels can be used to power cars both big and small. But full-sized solar cars aren't practical for most uses.

power cars. But with heavy panels and batteries to carry, the technology isn't practical for everyday cars.

Mayor Rush tells me a small, slow vehicle such as a golf cart can get enough power from a solar panel. And there are electric

A solar car prepares for the World Solar Challenge solar car race in Australia.

cars that work on batteries. But instead of relying on PVs on their roofs, these cars have to be recharged with electricity from the grid.

"PVs won't be powering our buses and trains any time soon, either," Mayor Rush says. But scientists have found ways to power large vehicles using solar power. She points me to an exhibit showing posters of solar boats, airplanes, and space vehicles.

"In 2007, a PV-powered boat crossed the Atlantic Ocean using only solar power," Mayor Rush says. "And the military has been testing solar-powered unmanned airplanes. One set a record by flying high in the atmosphere for over two weeks."

Mayor Rush reminds me that scientists are using the sun's energy to power objects far away from Earth as well. A solar-powered rover explored the surface of Mars for six years, even though about 44 percent less sunshine reaches Mars than Earth. A bit closer to home, Japanese scientists are working on an orbiting satellite that will beam enough solar energy back to Earth to power 294,000 homes.

I'm impressed. I wonder what other solar innovations are out there.

CHAPTER 8

Bright Ideas

My last stop is Nano Labs in Iowa. The company's president, Haddi Reyhani, is giving me a tour of the building.

As we start down the hall, Mr. Reyhani tells me, "Solar energy is still expensive, and businesses like mine have to make a profit. So labs like this one are testing ways to make it cheaper."

He shows me to a window that looks into a lab. Inside, researchers dressed in white uniforms and masks push buttons and turn knobs on machines that look like complicated microscopes.

Mr. Reyhani explains that I am looking at a clean room. Even one bit of dirt would ruin the work the researchers are doing. He tells me they are creating nanoparticles—tubes, fibers, and balls that can be smaller than specks of dust. Nanoparticles are so small that they are measured in nanometers—one billionth of a meter.

Nanoparticles can make solar cells more efficient. Nanoparticles make solar cells bumpy, giving them a larger surface area with which to absorb sunlight. A cell with certain types of nanoparticles can also absorb a broader range of radiation. That means it could turn infrared (nonvisible) light, as well as visible light, into electricity.

"Solar panels are usually made of heavy, costly silicon," says Mr. Reyhani. He hands me a small, rectangular piece of film and says, "But this is a solar panel made with nanoparticles on plastic. It's much cheaper and lighter than traditional solar cells, and it's even bendable."

Next we head outside to admire a wall of south-facing windows. "Here, we're testing thin-film silicon that can turn windows into

Researchers are working hard to make solar energy cheaper and more efficient.

THE ARTIFICIAL LEAF

Researchers at the Massachusetts Institute of Technology have been working to develop an artificial leaf. The "leaf" is actually a silicon solar cell about the size of a playing card. When dropped in a glass of water, solar energy on the leaf splits the water molecules into oxygen and hydrogen atoms. The hydrogen can be burned as a fuel that does not release carbon dioxide or other pollutants. But unfortunately the leaf is not very efficient. Only about 2.5 percent of the solar energy captured in the leaf can be turned into hydrogen. Researchers are working to make the technology more efficient. Other scientists are working to develop commercial cars that run solely on hydrogen. Someday, cars powered by artificial leaf–produced hydrogen could be driving around our city streets.

solar cells," Mr. Reyhani tells me. "Imagine a city of skyscrapers in which thousands of windows turn sunshine into electricity."

Mr. Reyhani says that buildings would not need to be rebuilt in order to use nanotechnology. At Nano Labs, researchers are working on developing films that could be retrofitted, or used on existing buildings.

But efficiency often improves if solar energy capture is part of a building's original design. "Planning homes, offices, and other buildings with PVs in mind is called building-integrated photovoltaics," says Mr. Reyhani. "Architects are now teaming with engineers in order to create designs that work well but aren't too expensive."

"We're running a business," he tells me. "Mathematicians on our staff figure out how much different nanofilms cost

Someday, huge cities like New York City might have skyscrapers with windows that can convert solar energy into electricity.

compared with how much electricity they make. If the cost of fossil fuels is low," he says, "electricity from our films is still too expensive for buyers to want them. But if coal and other fuel costs rise, then our system starts to save more and more."

CHAPTER 9

Your Turn

You've had a chance to follow Sage as she did her research. Now it's time to think about what you've learned. Solar power can't solve all of our energy problems right now, but it can help. To meet a larger share of our energy needs, solar power must be inexpensive and efficient enough to replace fossil fuels. Photovoltaic solar panels on homes and businesses are one option, but they are inefficient, and it can take many years for them to save enough electricity to be worth the cost of installing them. These individual solar panels can only provide solar power on a small scale. Big solar farms can provide solar power on a larger scale. However, these farms take up a lot of land, including former animal habitats. Researchers are working to improve solar power with new technology, such as nanoparticles. Soon, more and more of our energy may come from the sun.

What do you think? Is solar energy worth the expense?

YOU DECIDE

1. Do you think the pros of solar power outweigh the cons? Why or why not?
2. Do you think Sage should try to convince her school to invest in solar power? Why or why not?
3. What can you do to cut down on your energy use? Think about technology, such as solar panels, as well as ways to change your behavior, such as walking instead of driving.
4. Thinking about all you've learned about solar power, which technology do you think has the brightest future? Why?
5. How big of a role do you think solar power will play in our energy future? Why?

GLOSSARY

atoms: The smallest particles of elements.

carbon: An element common in living things that can trigger climate change in excess amounts.

carbon dioxide: A greenhouse gas released by the burning of fossil fuels.

climate change: Long-term changes in the earth's weather patterns.

concentrated solar power (CSP): A method of creating energy that uses lenses or mirrors to focus sunlight into a small beam, which boils water to create steam, turning a turbine that generates electricity.

current: The flow of electrical charge through a substance.

kilowatt-hour (kWh): The unit used to measure electrical energy use over time.

nanoparticles: Tiny particles that can be used to improve solar cells.

passive solar energy: Systems that absorb, use, and store solar energy to provide heat or light without relying on machines.

photosynthesis: A process through which carbon dioxide, water, and light energy are converted to energy for plants.

photovoltaic cells (PVs): Devices that convert sunlight into electrical current.

renewable: When something can be replaced by natural environmental cycles.

satellite: An object that travels around, or orbits, another object.

solar radiation: Energy given off by the sun.

thermal mass: The property of a material that allows it to absorb, store, and later release heat.

EXPLORE FURTHER

Pizza Box Solar Oven

Next time your family orders a pizza, save a pizza box. Cut the lid on three sides to make a rectangular flap. Cover the inside of the cardboard flap with aluminum foil with the shiny side out. Next, seal the opening with strong plastic wrap, using tape to secure it firmly. Finally, line the inside of the box with sturdy black paper. Now you can cook s'mores in the afternoon sun! Think about what you know about CSP. What does the aluminum foil do? Why use black paper instead of white?

Create Your Own Greenhouse Effect

Place two identical thermometers side by side on a sunny table outside or on a shelf by an open window. After five minutes, record their temperatures. Then place a glass jar over one thermometer. Record each temperature every minute for 10 minutes. Try the test again on another day with two jars, one lined with black construction paper and one lined with aluminum foil, shiny side out. Check the temperatures after 10 minutes. What components of our atmosphere act like the glass jar in your experiment, trapping heat? What absorbs heat, and what reflects it? Could scientists use this knowledge to help reduce climate change?

Shrink Your Carbon Footprint

The amount of greenhouse gases you produce is sometimes called your carbon footprint. Visit an online carbon footprint calculator to estimate how much carbon dioxide your household produces in a year. Examine your results—how can you reduce emissions? Can you hang laundry in the sun instead of running the clothes drier? Can you replace outdoor lighting with solar-powered lamps? What about growing your own food to reduce driving trips to the grocery store? What are some other things you could do to reduce your emissions?

Using the sun as a clothes dryer can help you cut down on your energy use.

SELECTED BIBLIOGRAPHY

Ball, Jeffrey. "The Homely Costs of Energy Conservation." *Wall Street Journal*, August 7, 2009. Web. Accessed July 18, 2012.

Biello, David. "Green Energy's Big Challenge: The Daunting Task of Scaling Up." *Yale Environment 360*, January 20, 2011. Web. Accessed July 18, 2012.

Cart, Julie. "Desert Tortoises Are Obstacles to Mojave Solar Development." *Washington Post*, March 18, 2012. Web. Accessed July 18, 2012.

Owen, David. "The Artificial Leaf: David Nocera's Vision for Sustainable Energy." *New Yorker Magazine*, May 14, 2012, 68–74. Print.

FURTHER INFORMATION

Books

Gardner, Robert. *Energy: Green Science Projects About Solar, Wind, and Water Power.* Berkeley Heights, NJ: Enslow Publishers, 2011.

Solway, Andrew. *Harnessing the Sun's Energy*. Chicago: Heinemann Library, 2009.

Spetgang, Tilly, and Malcolm Wells. *The Kids' Solar Energy Book Even Grown-Ups Can Understand*. Watertown, MA: Imagine Publishing, 2011.

Websites

http://www.alliantenergykids.com/energyandtheenvironment/renewableenergy/022400
Learn more about the history of solar power, and learn how to build a solar collector.

http://dsc.discovery.com/tags/solar-power/
Solar cell phones, solar film for windows, and other solar innovations are provided by the Discovery Channel's website.

http://www.eia.gov/kids/energy.cfm?page=solar_home-basics
This U.S. Energy Information Administration site provides world maps of solar resources, diagrams of solar power towers, and much more.

INDEX